HARVESTING RAINWATER HOMESTEAD

A Beginner's ultimate Guide to Build and Maintain Your Own Sustainable Clean Water System for Your Home, Rural Farm, and Homestead

Alex kava

Copyright

Table of Contents

Introduction

In the quaint town of Meadowville, nestled between rolling hills and lush greenery, lived a person named Alex. Known for their love of sustainable living and a deep connection to the land, Alex had always sought innovative ways to harmonize with nature. One day, while browsing the shelves of the town's cozy bookstore, Alex's eyes were drawn to a book with a title that sparked immediate curiosity – **"HARVESTING RAINWATER HOMESTEAD."**

Intrigued, Alex reached for the book, its cover adorned with vibrant homesteads. Little did Alex know that this simple act of pulling that book from the shelf would set in motion a transformative journey.

As Alex delved into the pages of **"HARVESTING RAINWATER HOMESTEAD,"** a world of possibilities unfolded. The book, written by a seasoned expert in sustainable living, became Alex's guide to unlocking the potential of rainwater harvesting. The chapters unfolded with insights into the science of rainwater, the art of planning a system, and the hands-on construction of the infrastructure.

What made this book particularly special was its ability to convey complex concepts in a way that resonated with Alex's passion for the environment. With each page turned, Alex envisioned a homestead where rainwater became a precious resource, sustaining gardens, quenching the thirst of livestock, and contributing to a more self-sufficient lifestyle.

The practical tips on maintaining the system, the importance of seasonal considerations, and the integration of rainwater harvesting with other sustainable practices resonated deeply with Alex's values. As Alex absorbed the knowledge, it wasn't just about building a rainwater harvesting system; it was about cultivating a mindful and eco-conscious way of life.

Remarkably, by the time Alex finished reading **"HARVESTING RAINWATER HOMESTEAD,"** a newfound confidence and excitement had taken root. Armed with the wisdom gained from the book, Alex set out to implement the lessons learned. The rooftops were transformed into efficient collection surfaces, storage tanks were strategically placed, and a first flush diverter became a key component of the system.

Neighbors soon noticed the changes in Alex's homestead – the flourishing gardens, the healthy-looking livestock, and the abundant water supply during dry spells. Intrigued, they inquired about the transformation. In response, Alex didn't just share the technical details; they passed along the book that had sparked this journey.

In Meadowville, a quiet revolution began. As more homesteads embraced rainwater harvesting, the town transformed into a community that valued sustainability and shared knowledge. All of this, sparked by the simple act of discovering a book with a compelling title – **"HARVESTING RAINWATER HOMESTEAD."**

As seasons changed and raindrops continued to fall, Meadowville became a testament to the power of a single book to inspire change. The person who once merely read about rainwater harvesting had become a catalyst for a ripple effect, demonstrating that sometimes, the key to a sustainable future lies in the wisdom found between the covers of a well-chosen book.

Chapter 1: Introduction

Welcome to "Harvesting Rainwater Homestead: A Beginner's Ultimate Guide to Build and Maintain Your Own Sustainable Clean Water System for Your Home, Rural Farm, and Homestead." In this comprehensive guide, we embark on a journey to explore the invaluable practice of rainwater harvesting and its transformative potential for homesteads, rural farms, and homes alike.

Water scarcity is an escalating global concern, and as stewards of our own living spaces, it becomes imperative to seek sustainable solutions. Rainwater harvesting stands out as a pragmatic and eco-friendly approach, offering a myriad of benefits beyond the conventional water sources.

Benefits for Homesteads and Rural Farms

Chapter 1 delves into the multifaceted advantages of integrating rainwater harvesting into your homestead or rural farm. From reducing dependence on municipal water supplies to providing a reliable source for irrigation, this section emphasizes the positive environmental impact and long-term cost savings associated with this practice.

We'll explore how harvesting rainwater not only conserves precious natural resources but also promotes resilience in the face of erratic weather patterns. The potential to contribute to a more sustainable future is in your hands, quite literally, as we guide you through the steps of setting up your own rainwater harvesting system.

Overview of the Guide

To kickstart your journey, this section outlines what you can expect from the guide, providing a roadmap for the chapters that follow. Each subsequent chapter builds on the knowledge gained, taking you from understanding the basics of rainwater to planning, constructing, and maintaining a robust harvesting system.

As you immerse yourself in the world of rainwater harvesting, we hope this guide inspires you to adopt and customize these practices for your unique homestead or farm. So, let's begin our exploration of sustainable water solutions, empowering you to create a more self-sufficient and eco-conscious living environment.

The Importance of Rainwater Harvesting

In the face of escalating environmental challenges, the practice of rainwater harvesting emerges as a beacon of sustainability, offering a crucial solution to the ever-growing global water crisis. This introductory chapter aims to illuminate the profound importance of rainwater harvesting, especially for those seeking a more self-reliant and eco-friendly approach to water consumption.

Preserving a Precious Resource

Water, the essence of life, is becoming an increasingly scarce commodity in many regions worldwide. As climate change brings about unpredictable weather patterns and exacerbates drought conditions, traditional water sources are strained to meet the demands of growing populations. In this context, rainwater harvesting stands as a reliable and environmentally conscious alternative.

A Sustainable Solution for Homesteads

Homesteads, rural farms, and homes can benefit immensely from harnessing the power of rainfall. Beyond being an accessible and cost-effective water source, rainwater harvesting offers a sustainable way to reduce reliance on external water supplies. This chapter explores the transformative potential of adopting rainwater harvesting practices, emphasizing the tangible advantages it brings to individual households and communities.

Environmental Stewardship in Action

Beyond the immediate benefits to the individual homestead, rainwater harvesting contributes to broader environmental conservation efforts. By capturing and utilizing rainwater, individuals become active stewards of their local ecosystems, lessening the strain on natural water bodies and promoting biodiversity. This chapter underscores the role of rainwater harvesting as a practical expression of environmental responsibility.

Addressing Water Scarcity Through Innovation

As we embark on this exploration of rainwater harvesting, we delve into the innovative techniques and technologies that make it an adaptable solution for diverse environments. From simple DIY setups to more sophisticated systems, this chapter sets the stage for understanding the varied approaches to rainwater harvesting and how they can be tailored to specific needs.

As we navigate through the subsequent chapters, you will gain insights into planning, building, and maintaining your rainwater harvesting system. Join us in this journey towards water sustainability, where every drop of rain becomes a valuable resource for a more resilient and environmentally conscious future.

Benefits for Homesteads and Rural Farms

As we delve into the heart of our guide, "Harvesting Rainwater Homestead: A Beginner's Ultimate Guide to Build and Maintain Your Own Sustainable Clean Water System for Your Home, Rural Farm, and Homestead," it's essential to unravel the profound benefits that rainwater harvesting brings to homesteads and rural farms.

A Resilient Water Source**

Homesteads and rural farms often grapple with the challenge of securing a reliable water supply, particularly in areas where access to municipal water systems may be limited or unreliable.

Rainwater harvesting emerges as a resilient and accessible solution, providing a consistent source of clean water independent of external factors.

Cost-Efficient Water Management

In an era where water bills can be a substantial portion of a household or farm's expenses, rainwater harvesting offers a cost-effective alternative. By capturing and utilizing rainwater, individuals can significantly reduce their reliance on expensive municipal water sources or energy-intensive well systems. This chapter explores the financial advantages of implementing a rainwater harvesting system for sustainable water management.

Irrigation Independence

For rural farms heavily reliant on irrigation, rainwater harvesting presents an opportunity for greater independence. The ability to collect and store rainwater for agricultural use not only conserves precious groundwater but also ensures a more consistent water supply during dry periods. We will explore the techniques and considerations for integrating rainwater into your irrigation practices, fostering a more sustainable and productive farming environment.

Environmental Conservation on Your Homestead

Beyond the immediate practical benefits, adopting rainwater harvesting practices transforms homesteads into hubs of environmental conservation. By lessening the demand on local water sources and reducing runoff, homesteads become active contributors to ecological health. This chapter underscores the role of rainwater harvesting in promoting sustainable land use and biodiversity on rural properties.

Community Empowerment Through Water Self-Sufficiency

In embracing rainwater harvesting, homesteads and rural farms become pioneers of water self-sufficiency. This not only enhances individual resilience but also contributes to community empowerment. This chapter explores the potential for shared water resources and the positive impact a collective commitment to rainwater harvesting can have on the broader community.

Join us on this exploration of the benefits that rainwater harvesting brings to homesteads and rural farms, as we pave the way toward a more sustainable, cost-efficient, and environmentally conscious approach to water management.

Overview of the Guide

In navigating the comprehensive realm of rainwater harvesting for homesteads, rural farms, and homes, it's essential to provide a clear roadmap for the journey that lies ahead. This chapter serves as a guide within our guide, offering an overview of the key themes and chapters that will unfold in this exploration of sustainable water solutions.

Understanding the Path Ahead

Before embarking on the practical aspects of rainwater harvesting, it's crucial to grasp the structure and purpose of this guide. We will outline the main sections, highlighting the progression from foundational knowledge to practical implementation, ensuring that each reader can approach the material with a clear understanding of what to expect.

Building a Foundation: The Basics of Rainwater

The initial chapters lay the groundwork by delving into the science behind rainwater, its unique properties, and how it distinguishes itself from conventional water sources. We'll explore the quantity and seasonal variations of rainfall, providing a foundational understanding that informs the subsequent stages of the guide.

Planning Your Water Future

Armed with a grasp of rainwater fundamentals, the guide shifts focus to the planning phase. Readers will be guided through assessing their homestead's water needs, selecting appropriate collection surfaces, and determining storage capacity. Legal and regulatory considerations will also be addressed, ensuring a comprehensive approach to the planning process.

Building the Blueprint: Construction and Installation

Once the planning is complete, attention turns to the practicalities of constructing a rainwater harvesting system. This section will detail the components of a basic system, offering insights into the decision-making process between DIY and professional installation. A step-by-step construction guide and troubleshooting tips will empower readers to bring their systems to life.

Sustaining the Flow: Maintenance and Upkeep

An often-overlooked aspect of rainwater harvesting is the ongoing maintenance required for a system's longevity. This section will explore the importance of regular inspections, cleaning routines, and strategies for winterization. Readers will gain valuable insights into repairing and upgrading their systems to ensure continued efficiency.

Maximizing Impact: Innovative Uses and Community Engagement

The final chapters elevate the discussion by exploring innovative uses for harvested rainwater and strategies for integrating rainwater harvesting with other sustainable practices. The guide concludes by emphasizing the potential for community education and engagement, empowering readers to become advocates for water sustainability in their localities.

As we embark on this journey through the chapters, each step is designed to empower you to build and maintain a rainwater harvesting system tailored to your unique needs. Join us in this exploration of sustainable water solutions, where theory meets practice, and every raindrop becomes a valuable resource for a more resilient and eco-conscious future.

Chapter 2: Understanding Rainwater

Rainwater, a natural gift from the skies, holds untapped potential as a sustainable water source for your homestead. In this chapter, we embark on a journey to deepen our comprehension of rainwater — its composition, unique characteristics, and the science that underpins its role in the broader context of water conservation.

The Science of Rainwater

To harness the benefits of rainwater harvesting, it's imperative to understand the fundamental science behind rainfall. We'll explore the atmospheric processes that lead to precipitation, providing insights into how water vapor condenses into droplets, ultimately forming the rain that nourishes our land. This foundational knowledge sets the stage for appreciating the purity and quality of rainwater.

How Rainwater Differs from Other Water Sources

Rainwater possesses distinct qualities that set it apart from traditional water sources. This section compares and contrasts rainwater with other forms of water, including tap water and well water. Understanding these differences is crucial for tailoring your rainwater harvesting system to maximize efficiency and address specific needs unique to your homestead or rural farm.

Quantity and Seasonal Variations

Rainfall patterns vary widely based on geographical location and climatic conditions. We delve into the factors influencing the quantity of rainfall, offering insights into the seasonal variations that impact harvesting potential. This knowledge forms the basis for effective system planning, ensuring your rainwater harvesting system aligns with the specific characteristics of the local climate.

As we unravel the intricacies of rainwater, you'll gain a deeper appreciation for this abundant resource and its potential to sustainably meet your water needs. Join us in the exploration of the science behind rainwater, laying the groundwork for the practical steps that follow in planning, constructing, and maintaining your very own rainwater harvesting system.

The Science of Rainwater

Rainwater, seemingly simple droplets falling from the sky, is a complex product of atmospheric processes that play a pivotal role in sustaining life on Earth. This section delves into the intricate science behind rainwater, unraveling the journey of water vapor to liquid form.

Condensation in the Atmosphere

The process begins high above us, where water vapor condenses into tiny droplets in the atmosphere. As warm air rises, it cools, causing water vapor to condense around particles in the air. These droplets combine to form clouds, and when they reach a critical mass, gravity takes over, and rain falls to the Earth.

The Purity of Rainwater

Unlike water from other sources, rainwater starts in a relatively pure state. It is free from the contaminants often found in ground or surface water. However, as it travels from the sky to collection surfaces and storage, it can pick up impurities. Understanding this natural purity is key to appreciating the potential for harvesting clean, high-quality water.

Chemical Composition and Benefits

Rainwater is a soft water, naturally low in minerals like calcium and magnesium. This softness makes it suitable for various household uses, and its lack of treatment chemicals makes it a compelling choice for irrigation. Exploring the chemical composition of rainwater enhances our understanding of its versatility and potential applications on a homestead or farm.

Factors Influencing Rainfall

To grasp the availability of rainwater in a specific region, it's crucial to understand the factors influencing rainfall. These include prevailing wind patterns, topography, and proximity to water bodies. By comprehending these factors, homesteaders can better predict the quantity and frequency of rainfall, aiding in the efficient design of a rainwater harvesting system.

Sustainable Water Cycle Participation

Rainwater harvesting is not just about collecting water; it's about participating in the planet's natural water cycle. By capturing rainwater, we contribute to a sustainable water management system, reducing reliance on traditional sources and mitigating the impact of droughts and water scarcity.

As we navigate the intricacies of the science behind rainwater, a profound appreciation for its purity and sustainability emerges. Join us in this exploration, laying the groundwork for the practical steps that follow in planning, constructing, and maintaining your personalized rainwater harvesting system.

How Rainwater Differs from Other Water Sources

Rainwater, as a source of sustenance for your homestead, stands apart from conventional water sources in its composition and characteristics. In this section, we delve into the distinctive qualities that differentiate rainwater from other water supplies, shedding light on why it is a unique and valuable resource.

Purity Unveiled: Rainwater vs. Tap Water

Unlike tap water, often treated with chemicals for purification, rainwater begins its journey in the atmosphere, untouched by human-made additives. This inherent purity makes rainwater an appealing choice for various uses, from household consumption to gardening. Understanding this distinction forms the foundation for harnessing rainwater's natural benefits.

Softness and Mineral Content

Rainwater is inherently soft, meaning it contains fewer minerals like calcium and magnesium compared to hard water from wells or municipal supplies. This softness not only contributes to a more pleasant sensory experience but also has practical implications for household appliances and irrigation systems. Exploring the differences in mineral content helps tailor the use of rainwater to specific needs.

Freedom from Groundwater Impurities

Groundwater, while a valuable resource, can contain impurities from the soil it passes through. Rainwater, falling directly from the sky, avoids these potential contaminants. Examining this distinction underscores the potential for rainwater harvesting to provide a cleaner and more sustainable alternative, particularly in regions where groundwater quality may be a concern.

Collecting Nature's Distillation: Rainwater vs. Surface Water

Rainwater, essentially nature's distillation process, differs significantly from surface water found in lakes or rivers. While surface water can carry sediment and pollutants, rainwater starts as a relatively pure substance. This section explores how harvesting rainwater minimizes the need for extensive filtration and treatment, making it an attractive option for those seeking a more natural water source.

Adaptability and Versatility

Understanding how rainwater differs from other sources is essential for optimizing its use on a homestead. From reducing reliance on municipal water to providing an alternative for irrigation, rainwater's adaptability emerges as a key advantage. Exploring these differences empowers homesteaders to make informed decisions about integrating rainwater into their daily water consumption and management practices.

As we unravel the unique qualities of rainwater, we pave the way for harnessing its potential on your homestead. Join us in this exploration, where the distinctive attributes of rainwater become the foundation for designing a sustainable and self-sufficient water system.

Quantity and Seasonal Variations

To harness the full potential of rainwater for your homestead, it's crucial to grasp the nuances of its quantity and the seasonal variations that influence its availability. In this section, we explore the factors that impact the volume of rainfall, providing insights into how understanding these patterns is instrumental in planning an effective rainwater harvesting system.

Factors Influencing Rainfall Quantity

The quantity of rainwater available for harvesting is influenced by a multitude of factors. Geographic location, prevailing wind patterns, and topographical features all play integral roles. Coastal areas might experience more consistent rainfall due to oceanic influences, while mountainous regions could witness orographic rainfall as moist air is lifted over elevated terrain. By comprehending these factors, homesteaders can anticipate the potential yield of rainwater in their specific locale.

Seasonal Dynamics and Harvesting Potential

Understanding the seasonal variations in rainfall is paramount for optimizing the design and efficiency of a rainwater harvesting system. Some regions experience distinct wet and dry seasons, while others may have more evenly distributed rainfall throughout the year. This section delves into the implications of these variations, guiding homesteaders on how to adapt their systems to capture and store rainwater effectively during both abundance and scarcity.

Planning for Dry Spells: The Importance of Storage Capacity

In regions where dry spells are common, having a sufficient storage capacity becomes imperative. This involves not only understanding the average rainfall but also planning for scenarios where rainfall may be scarce for extended periods. By exploring storage solutions that align with the local climate, homesteaders can ensure a continuous and reliable water supply even during dry seasons.

Harvesting Rainwater in Urban and Rural Contexts

Urban and rural environments often present distinct challenges and opportunities for rainwater harvesting. Urban areas may have limited space for large storage tanks, necessitating innovative solutions. Conversely, rural farms might have more expansive areas for collection but face logistical challenges. This section addresses the considerations for both settings, ensuring that the principles discussed can be applied adaptively.

Climate Change Resilience and Adaptation

Given the evolving patterns of climate change, understanding the historical and projected shifts in rainfall becomes integral. Homesteaders equipped with this knowledge can adapt their rainwater harvesting systems to anticipate changes in precipitation patterns, contributing to increased resilience in the face of a changing climate.

As we delve into the quantity and seasonal variations of rainwater, we lay the groundwork for a strategic and adaptive approach to harvesting. Join us in this exploration, where the ebb and flow of rainfall become the guiding factors in designing a resilient and sustainable rainwater harvesting system for your homestead.

Chapter 3: Planning Your Rainwater Harvesting System

As we embark on the practical journey of implementing a rainwater harvesting system, meticulous planning becomes the cornerstone of success. In this chapter, we guide you through the essential steps of assessing your homestead's water needs, selecting appropriate collection surfaces, determining storage capacity, and navigating legal and regulatory considerations.

Assessing Your Homestead's Water Needs

Before constructing your rainwater harvesting system, a thorough understanding of your water requirements is paramount. This involves evaluating daily water usage for household activities, irrigation needs for gardens or crops, and any livestock or farming operations. This section provides a comprehensive guide to conducting a water needs assessment, ensuring that your rainwater harvesting system is tailored to meet your specific demands.

Choosing the Right Collection Surfaces

Effective rainwater harvesting begins with selecting suitable collection surfaces. This section explores various options, from traditional roofs to specialized surfaces designed to maximize water capture. Factors such as surface material, area, and cleanliness play crucial roles in determining the efficiency of your system. We delve into the considerations for choosing the most appropriate collection surfaces based on your homestead's layout and requirements.

Determining Storage Capacity

The storage capacity of your rainwater harvesting system is a key determinant of its functionality. This section guides you through the calculations needed to determine the optimal size for your storage tanks. By considering factors like average rainfall, the size of your catchment area, and your homestead's water consumption patterns, you can ensure that your system provides a reliable and consistent water supply.

Legal and Regulatory Considerations

Navigating legal and regulatory considerations is a crucial aspect of planning your rainwater harvesting system. Different regions may have specific rules governing water rights, collection practices, and the installation of storage tanks. This section provides insights into researching and complying with local regulations, ensuring that your system is not only efficient but also legally sound.

Environmental Impact Assessment

Additionally, we explore the environmental impact of your rainwater harvesting system, emphasizing the importance of minimizing any potential negative effects on the surrounding ecosystem. By conducting an environmental impact assessment, you can proactively address concerns and contribute to the sustainability of your homestead.

As you embark on the planning phase, meticulous consideration of these factors will lay the groundwork for a successful and sustainable rainwater harvesting system. Join us in this chapter as we guide you through the essential steps to ensure your system is custom-tailored to meet the unique needs of your homestead.

Assessing Your Homestead's Water Needs

Before the first raindrop is captured, a crucial step in planning your rainwater harvesting system is a comprehensive assessment of your homestead's water needs. This foundational process ensures that your system is tailored to meet the specific demands of your household, garden, and any agricultural activities.

Understanding Daily Water Usage

Begin by evaluating the daily water usage for your household. Consider activities such as bathing, cooking, cleaning, and any additional water-dependent tasks. This assessment provides a baseline for determining the minimum water volume required to sustain your daily activities.

Irrigation and Agricultural Demands

For homesteads with gardens, crops, or livestock, an in-depth evaluation of irrigation and agricultural water needs is essential. Consider the type and size of crops, the frequency of irrigation, and the water requirements of your livestock. This ensures that your rainwater harvesting system not only caters to household demands but also supports the growth and well-being of your plants and animals.

Forecasting Future Water Needs

Anticipate future changes in water consumption. If you plan to expand your homestead, increase agricultural activities, or add new water-dependent elements, factor these considerations into your assessment. Planning for future needs ensures the scalability and longevity of your rainwater harvesting system.

Conservation Measures**

As part of your assessment, identify opportunities for water conservation. Implementing water-saving fixtures, practicing mindful water use, and investing in efficient appliances can significantly impact the overall water demand. By incorporating conservation measures into your plan, you not only optimize your system but also contribute to a sustainable water management strategy.

Seasonal Variations and Peak Demand

Consider the impact of seasonal variations on water demand. In certain seasons, you may experience increased irrigation requirements or additional water needs for special projects. Understanding these fluctuations allows you to dimension your system to handle peak demand periods effectively.

Community Engagement

For those in communal living situations or tight-knit neighborhoods, involving the community in the water needs assessment can lead to collective solutions. Shared rainwater harvesting systems and coordinated conservation efforts can enhance water resilience for the entire community.

By thoroughly assessing your homestead's water needs, you lay the groundwork for a rainwater harvesting system that is not only efficient but also responsive to the unique demands of your lifestyle and environment. Join us in the next sections as we explore the selection of appropriate collection surfaces, determining storage capacity, and addressing legal and regulatory considerations in crafting a holistic plan for your rainwater harvesting system.

Choosing the Right Collection Surfaces

Selecting appropriate collection surfaces is a pivotal step in the planning process of your rainwater harvesting system. The efficiency and effectiveness of your system hinge on the types of surfaces from which rainwater can be captured. This section guides you through considerations for choosing the right collection surfaces tailored to your homestead's layout and water needs.

Surface Material and Cleanliness

The material of your collection surfaces significantly impacts the quality of the harvested rainwater. Rooftops made of materials such as metal, tile, or asphalt shingles are commonly used for rainwater collection due to their durability and smooth surfaces. Avoid materials that may introduce contaminants or impurities into the collected water. Additionally, ensuring the cleanliness of these surfaces is crucial for maintaining water quality.

Catchment Area Size

The size of your catchment area directly influences the volume of rainwater you can harvest. Larger catchment areas, such as expansive roofs or open spaces, yield more water. Evaluate the available space on your homestead, considering roofs, driveways, or other impermeable surfaces where rainwater can be effectively collected.

Consideration for Specialized Collection Surfaces

In some cases, specialized collection surfaces can be employed to enhance water capture efficiency. This might include installing purpose-built structures like rain gardens, green roofs, or permeable pavements. These surfaces are designed to maximize water absorption and reduce runoff, offering environmentally friendly alternatives for rainwater collection.

Maintenance and Accessibility

Consider the maintenance requirements of your chosen collection surfaces. Ensure that they are easily accessible for routine cleaning to prevent the buildup of debris, dust, or pollutants. Regular maintenance not only safeguards water quality but also prolongs the life of your rainwater harvesting system.

Adaptability to Local Climate

The local climate plays a role in determining the effectiveness of different collection surfaces. For areas with heavy snowfall, a smooth, steep roof may facilitate snow shedding, preventing the accumulation of excess weight. Conversely, in regions with frequent rainfall, surfaces that minimize splashing and facilitate efficient runoff may be more suitable.

Integration with Aesthetic and Functional Design

The choice of collection surfaces can also be an opportunity to integrate aesthetic and functional design elements into your homestead. Explore options that align with the overall design and purpose of your property, ensuring that the rainwater harvesting system complements rather than detracts from the visual appeal of your surroundings.

As you navigate the process of selecting collection surfaces, you pave the way for efficient rainwater capture that aligns with the unique characteristics of your homestead. Join us in the subsequent sections as we delve into determining storage capacity and addressing legal and regulatory considerations, bringing you closer to the realization of a robust rainwater harvesting system.

Determining Storage Capacity

After assessing your homestead's water needs and selecting appropriate collection surfaces, the next critical step in planning your rainwater harvesting system is determining the storage capacity.

Adequate storage ensures a reliable and consistent water supply, especially during dry spells or periods of minimal rainfall. This section provides guidance on calculating and optimizing the storage capacity for your unique requirements.

Calculating Average Rainfall and Catchment Area

Begin by calculating the average rainfall in your area. This information, combined with the size of your catchment area, forms the basis for estimating the potential volume of rainwater you can harvest. Meteorological data, local weather patterns, and historical rainfall records can aid in making informed calculations.

Assessing Water Demand and Usage Patterns

Align the determined rainfall potential with your homestead's water needs. Consider the assessed water demand for household activities, irrigation, and any agricultural purposes. Factor in the usage patterns and distribution of water consumption throughout the year, accounting for peak demand periods.

Planning for Dry Spells and Seasonal Variations

Storage capacity must account for dry spells and seasonal variations in rainfall. During periods of limited precipitation, having sufficient reserves becomes critical. Consider the length of dry spells in your region and plan for storage that ensures a consistent water supply even when rainfall is scarce.

Determining Tank Size and Quantity

Based on the calculated water needs and potential harvestable volume, determine the size and quantity of storage tanks required. The choice of tank material, whether plastic, concrete, or other materials, can impact durability, cost, and maintenance. Explore tank options that align with your storage capacity goals and local climate conditions.

Scaling for Future Expansion**

Anticipate future needs and expansion of your homestead. A rainwater harvesting system is a long-term investment, and planning for future growth ensures that your system remains scalable. This may involve installing additional tanks or designing the system to accommodate increased water demand.

Considering Redundancy and Backup Systems

Integrate redundancy and backup systems into your storage planning. This involves having alternative water sources or backup storage options in case of unforeseen circumstances such as equipment failure or extreme weather events. Redundancy enhances the resilience of your rainwater harvesting system.

Evaluating Space Constraints and Aesthetics

Assess the available space on your homestead for installing storage tanks. Consider the aesthetics and visual impact of the tanks, especially in residential areas. Explore creative solutions such as underground or concealed tanks that minimize visual impact while maximizing storage capacity.

By meticulously determining storage capacity, you lay the foundation for a rainwater harvesting system that meets your homestead's water needs sustainably. Join us in the upcoming sections as we address legal and regulatory considerations, ensuring that your system not only functions efficiently but also complies with local guidelines and regulations.

Legal and Regulatory Considerations

Navigating legal and regulatory considerations is a crucial aspect of planning your rainwater harvesting system. Compliance with local laws ensures that your system is not only efficient but also adheres to environmental, safety, and water management guidelines. This section provides insights into researching, understanding, and addressing the legal and regulatory landscape for rainwater harvesting in your specific region.

Researching Local Regulations

Initiate the planning process by researching local regulations pertaining to rainwater harvesting. Regulations can vary significantly from one region to another, influencing aspects such as water rights, collection methods, storage capacities, and permissible uses. Consult local government offices, water authorities, or environmental agencies to gather comprehensive information.

Understanding Water Rights and Ownership

In some areas, water rights and ownership are legally defined. Understanding the legal framework regarding water rights ensures that your rainwater harvesting activities do not infringe upon established norms. In riparian or prior appropriation systems, compliance with these legal structures is particularly important.

Complying with Collection and Storage Guidelines

Local regulations may prescribe guidelines for the collection and storage of rainwater. These guidelines could include restrictions on specific collection surfaces, the use of certain materials for storage tanks, or limitations on storage capacities. Ensure that your planned system aligns with these guidelines to prevent legal complications.

Permitting and Approval Processes

Certain jurisdictions may require permits or approvals for rainwater harvesting systems. Familiarize yourself with the permitting process, including the documentation and submissions needed for approval. Understanding the administrative requirements streamlines the process of obtaining necessary permits before system installation.

Addressing Health and Safety Standards

Some regions may have health and safety standards governing the use of harvested rainwater, particularly for potable water purposes. Ensure that your system complies with these standards, and if needed, incorporate appropriate filtration or treatment methods to meet water quality requirements.

Considering Community or Homeowner Association Rules

If you reside in a community or belong to a homeowner association, there may be additional rules and covenants to consider. These may relate to the visual impact of the system, noise during installation, or other community-specific concerns. Engage with the relevant authorities or associations to ensure harmony with community guidelines.

Environmental Impact Assessment

Conduct an environmental impact assessment as part of your planning process. This involves evaluating the potential ecological effects of your rainwater harvesting system, such as changes to local water tables or impacts on flora and fauna. Proactively addressing environmental concerns contributes to the sustainability of your system.

Educating and Collaborating with Local Authorities

Engage in proactive communication with local authorities. Educate them about the benefits of rainwater harvesting and seek guidance on best practices. Building a collaborative relationship with local agencies fosters understanding and may lead to supportive policies that encourage sustainable water management.

By navigating legal and regulatory considerations with diligence and compliance, you not only ensure the legality of your rainwater harvesting system but also contribute to a broader culture of responsible and sustainable water use. Join us in the upcoming sections as we delve into the practical aspects of building your rainwater harvesting system, transforming plans into a tangible and efficient reality.

Chapter 4: Building Your Rainwater Harvesting System

With a solid plan in place, it's time to transition from theory to practice and embark on the construction of your rainwater harvesting system. This chapter guides you through the practical steps of materializing your vision, from assembling the necessary components to ensuring the efficient installation and functionality of your system.

Gathering Materials and Components

Rainwater Collection Surfaces

Begin by ensuring that your chosen collection surfaces are in good condition and free of contaminants. Clean roofs, gutters, and other catchment areas thoroughly to prevent debris from entering the system.

Storage Tanks

Acquire the storage tanks in accordance with your determined capacity. Consider factors such as tank material, size, and quantity. Install tanks on stable foundations and ensure that they are properly secured.

Gutters and Downspouts

Ensure that gutters and downspouts are securely attached to your collection surfaces. Regularly clean and inspect them to prevent clogs and optimize water flow into the storage tanks.

First Flush Diverters

Integrate first flush diverters into your system to redirect the initial rainwater flow, which may contain debris, pollutants, or contaminants. This helps improve the overall quality of the collected water.

Filtration Systems

Depending on your water usage needs, incorporate appropriate filtration systems. This may include mesh filters, sediment filters, or more advanced filtration methods to enhance the purity of the harvested rainwater.

Constructing the Conveyance System

Pipes and Conduits

Install pipes and conduits to convey rainwater from the collection surfaces to the storage tanks. Ensure that these conduits are correctly sloped for efficient water flow and that they are securely attached to prevent leaks.

First Flush and Filtration Integration

Integrate the first flush diverters and filtration systems into the conveyance system. These components play a crucial role in maintaining water quality and preventing contaminants from entering the storage tanks.

Overflow Systems

Incorporate overflow systems to prevent excessive water from damaging the storage tanks or surrounding areas. This ensures a controlled release of excess water during heavy rainfall.

Implementing the Storage System

Tank Installation

Position and install the storage tanks in accordance with your planned layout. Ensure that the tanks are level and securely anchored to prevent shifting or damage.

Inlet and Outlet Connections

Establish inlet and outlet connections to facilitate the seamless flow of water into and out of the storage tanks. Confirm that connections are watertight and that there is no risk of contamination.

Water Level Indicators

Integrate water level indicators to monitor the volume of water in your tanks accurately. This helps you manage your water reserves effectively and plan for periods of high or low rainfall.

Testing and Troubleshooting

System Testing

Conduct thorough testing of your rainwater harvesting system to ensure that all components are functioning as intended. Verify that water flows smoothly from collection surfaces to storage tanks and that filtration systems are effective.

Troubleshooting

Identify and address any issues that arise during testing promptly. Common challenges may include clogged filters, leaks, or inefficient water flow. Troubleshoot these problems to optimize the performance of your system.

Regular Maintenance and Upkeep

Scheduled Inspections

Establish a schedule for regular inspections and maintenance. This includes checking collection surfaces, cleaning gutters, inspecting filtration systems, and ensuring that storage tanks are in good condition.

Winterization Measures

Implement winterization measures if you reside in a region with cold temperatures. This may involve draining the system, protecting pipes from freezing, and safeguarding storage tanks from potential damage.

Community Engagement and Education

Educational Initiatives

Engage with your community to share knowledge about rainwater harvesting benefits and practices.

Foster a sense of environmental responsibility and encourage others to adopt sustainable water management practices.

Collaboration with Local Authorities

Maintain open communication with local authorities. Share information about your rainwater harvesting system, addressing any concerns they may have and contributing to a positive dialogue on sustainable water practices.

By methodically building and maintaining your rainwater harvesting system, you transform theoretical plans into a tangible, efficient, and sustainable water solution for your homestead. Join us in the subsequent sections as we explore innovative uses for harvested rainwater and strategies for maximizing its impact on your property.

Components of a Basic System

Embarking on the construction of your rainwater harvesting system involves assembling essential components to ensure efficient collection, conveyance, and storage of rainwater. In this section, we delve into the key components of a basic rainwater harvesting system, guiding you through the selection and installation process.

1. Rainwater Collection Surfaces:

Select appropriate surfaces for rainwater collection, typically roofs made of materials like metal, tile, or asphalt shingles. Ensure these surfaces are clean and well-maintained to prevent debris from entering the system.

2. Gutters and Downspouts:

Install gutters along the edges of your collection surfaces to channel rainwater. Downspouts guide the water from the gutters to the storage tanks. Regular cleaning and maintenance are crucial to prevent clogs.

3. First Flush Diverters:

Integrate first flush diverters to redirect the initial flow of rainwater away from the storage tanks. This helps eliminate debris, pollutants, and contaminants present on the collection surfaces, improving water quality.

4. Filtration Systems:

Depending on your water usage needs, incorporate filtration systems. This may include mesh filters, sediment filters, or more advanced filtration methods to enhance the purity of the harvested rainwater.

5. Pipes and Conduits:

Install pipes and conduits to convey rainwater from the collection surfaces to the storage tanks. Ensure correct slope for efficient water flow and secure connections to prevent leaks.

6. Storage Tanks:

Select storage tanks based on your determined capacity, considering factors such as material, size, and quantity. Install tanks on stable foundations, ensuring they are level and securely anchored.

7. Overflow Systems:

Incorporate overflow systems to manage excess water during heavy rainfall. This prevents damage to the storage tanks or surrounding areas and ensures controlled release of excess water.

8. Inlet and Outlet Connections:

Establish inlet and outlet connections for seamless water flow into and out of the storage tanks. Confirm watertight connections to prevent contamination.

9. Water Level Indicators:

Integrate water level indicators to monitor tank volumes accurately. This aids in managing water reserves effectively and planning for periods of high or low rainfall.

Assembling these fundamental components forms the basis of a basic rainwater harvesting system. In the subsequent sections, we will guide you through the construction process, testing, troubleshooting, and ongoing maintenance, ensuring a robust and sustainable water solution for your homestead.

DIY vs. Professional Installation

The decision to embark on a do-it-yourself (DIY) installation or opt for professional assistance in building your rainwater harvesting system hinges on various factors, including your level of expertise, available time, and the complexity of the planned system. In this section, we explore the considerations for both DIY and professional installations to help you make an informed choice.

DIY Installation:

Pros:

1. Cost Savings: DIY installations often come with lower upfront costs, as you can purchase materials and components independently.

2. Personalization: You have the flexibility to design and customize the system according to your specific needs, preferences, and available space.

3. Learning Experience: Undertaking a DIY project provides valuable hands-on experience, enhancing your understanding of the system's functionality.

4. Sense of Accomplishment: Completing the installation yourself can be rewarding and instill a sense of accomplishment.

Cons:

1. Skill Requirement: DIY installations may require a certain level of technical skill, especially in areas such as plumbing, construction, and system design.

2. Time-Consuming: Building the system yourself can be time-consuming, particularly if you are learning as you go and troubleshooting unforeseen challenges.

3. Potential Errors: Inexperienced DIYers may encounter errors in system design or installation, which can impact the efficiency and effectiveness of the rainwater harvesting system.

Professional Installation:

Pros:

1. Expertise: Professionals bring specialized knowledge and expertise, ensuring that the system is designed and installed correctly.

2. Time-Efficient: Professional installations are typically quicker, as experienced contractors can efficiently navigate the process.

3. Warranty and Guarantees: Many professional installations come with warranties and guarantees, providing assurance and recourse in case of issues.

4. Compliance: Professionals are well-versed in local regulations, ensuring that the installation complies with all legal requirements.

Cons:

1. Higher Cost: Professional installations may have higher upfront costs due to labor, expertise, and potentially premium-quality materials.

2. Less Personalization: While professionals can tailor the system to your needs, there may be less room for personalization compared to a DIY approach.

3. Dependency: You rely on the availability and schedule of professionals, which may lead to longer wait times for installation.

Considerations:

1. Complexity of the System: A straightforward rainwater harvesting system may be suitable for a DIY approach, while more complex systems may benefit from professional expertise.

2. Budget Constraints: Assess your budget and weigh the cost implications of a DIY project against the potential long-term benefits of a professionally installed, optimized system.

3. Time Availability: Consider your availability and time constraints. If time is a critical factor, a professional installation may be more suitable.

4. Learning Curve: Evaluate your comfort level with the technical aspects of the installation. If you are confident in your skills or willing to learn, a DIY approach could be fulfilling.

5. Local Regulations: Familiarize yourself with local regulations. If compliance is complex, a professional can navigate these requirements more effectively.

Ultimately, the choice between DIY and professional installation depends on your individual circumstances and preferences. Whichever path you choose, careful planning and attention to detail will contribute to the success and efficiency of your rainwater harvesting system.

Step-by-Step Construction Guide

Embarking on the construction of your rainwater harvesting system involves a series of carefully orchestrated steps to ensure optimal efficiency and functionality. This step-by-step guide provides a comprehensive roadmap for turning your plans into a tangible and sustainable water solution for your homestead.

Step 1: Prepare the Collection Surfaces

Ensure that your chosen collection surfaces, typically the roof, are clean and free from debris. Regular maintenance of these surfaces is crucial for preventing contaminants from entering the system.

Step 2: Install Gutters and Downspouts

Attach gutters along the edges of your collection surfaces to channel rainwater effectively. Install downspouts to guide the water flow from the gutters to the storage tanks. Ensure secure connections and proper alignment.

Step 3: Integrate First Flush Diverter

Incorporate first flush diverters into the system to redirect the initial flow of rainwater away from the storage tanks. This diverts the first flush, containing debris and pollutants, ensuring cleaner water enters the storage tanks.

Step 4: Install Filtration Systems

Depending on your water usage needs, integrate filtration systems into the conveyance system. This may include mesh filters, sediment filters, or advanced filtration methods to enhance the purity of the harvested rainwater.

Step 5: Construct the Conveyance System

Install pipes and conduits to convey rainwater from the collection surfaces to the storage tanks. Ensure proper slope for efficient water flow and secure connections to prevent leaks. Integrate first flush diverters and filtration systems into the conveyance system.

Step 6: Install Storage Tanks

Position and install the storage tanks based on your determined capacity. Ensure tanks are level, securely anchored, and have proper foundations.

Establish inlet and outlet connections for seamless water flow.

Step 7: Incorporate Overflow Systems

Integrate overflow systems to manage excess water during heavy rainfall. This prevents damage to the storage tanks or surrounding areas and ensures controlled release of excess water.

Step 8: Implement Water Level Indicators

Integrate water level indicators to monitor tank volumes accurately. This aids in managing water reserves effectively and planning for periods of high or low rainfall.

Step 9: Test the System

Conduct thorough testing of the entire rainwater harvesting system. Verify that water flows smoothly from collection surfaces to storage tanks and that filtration systems are effective. Identify and address any issues that arise during testing.

Step 10: Troubleshoot and Optimize

Identify and troubleshoot any issues that may arise during testing. Address problems promptly to optimize the performance of your rainwater harvesting system.

Step 11: Regular Maintenance and Upkeep

Establish a schedule for regular inspections and maintenance. This includes checking collection surfaces, cleaning gutters, inspecting filtration systems, and ensuring that storage tanks are in good condition.

Step 12: Community Engagement and Education

Engage with your community to share knowledge about rainwater harvesting benefits and practices. Foster a sense of environmental responsibility and encourage others to adopt sustainable water management practices.

By following this step-by-step construction guide, you can methodically build a rainwater harvesting system that aligns with your homestead's unique needs. The journey from planning to implementation ensures not only a sustainable water solution but also a positive impact on your community and the environment.

Troubleshooting Common Issues

As with any system, rainwater harvesting systems may encounter issues that affect their efficiency and performance. Being able to troubleshoot common problems is crucial for maintaining a reliable and sustainable water solution for your homestead. In this section, we address potential issues and provide troubleshooting steps to keep your rainwater harvesting system operating optimally.

1. Clogged Gutters and Downspouts:

Issue: Reduced water flow due to debris accumulation in gutters and downspouts.

Troubleshooting:
- Regularly clean gutters and downspouts to remove leaves, twigs, and other debris.
- Install gutter guards to prevent large debris from entering the system.

2. Inefficient Water Flow:

Issue: Water not flowing smoothly through pipes and conduits.

Troubleshooting:
- Check for blockages in pipes and conduits and clear any debris.
- Ensure the correct slope for conduits to facilitate efficient water flow.
- Inspect and clean filters to prevent clogs.

3. Inadequate Water Storage:

Issue: Insufficient water reserves in storage tanks.

Troubleshooting:
- Verify that the collection surfaces are effectively directing water to the storage tanks.
- Confirm that the inlet and outlet connections are secure and watertight.
- Evaluate if the storage capacity matches the anticipated water needs.

4. Overflow Issues:

Issue: Overflowing storage tanks during heavy rainfall.

Troubleshooting:
- Integrate larger overflow systems or divert excess water to a designated area.
- Check that overflow pipes are unobstructed and functioning correctly.

5. Contaminated Water:

Issue: Poor water quality due to contaminants or pollutants.

Troubleshooting:
- Inspect first flush diverters and ensure they are diverting initial runoff effectively.
- Regularly clean and maintain filtration systems.
- Address potential sources of contamination in the collection surfaces.

6. Tank Leaks:

Issue: Water leakage from storage tanks.

Troubleshooting:
- Inspect tanks for visible cracks or damage and repair as needed.
- Check the integrity of inlet and outlet connections for leaks.
- Ensure tanks are properly anchored and have stable foundations.

7. Inaccurate Water Level Indicators:

Issue: Water level indicators providing inaccurate readings.

Troubleshooting:
- Calibrate water level indicators to ensure accuracy.
- Check for any obstructions or damage to the indicators.

8. Frozen Conduits in Cold Weather:

Issue: Freezing of pipes and conduits in cold temperatures.

Troubleshooting:
- Implement winterization measures, such as draining the system before freezing conditions.
- Insulate pipes to prevent freezing.

9. System Inactivity:

Issue: Lack of rainfall leading to system inactivity.

Troubleshooting:

- Implement alternative water sources during dry spells.
- Consider supplemental irrigation methods during periods of low rainfall.

10. Lack of Community Engagement:

Issue: Limited community awareness and engagement.

Troubleshooting:
- Initiate educational initiatives to inform the community about the benefits of rainwater harvesting.
- Foster open communication with neighbors and local authorities to address concerns and build support.

By promptly addressing and troubleshooting these common issues, you can ensure the continued functionality and effectiveness of your rainwater harvesting system. Regular maintenance, proactive monitoring, and community engagement contribute to the long-term success of your sustainable water solution.

Chapter 5: Maintenance and Upkeep

Maintaining and regularly inspecting your rainwater harvesting system is essential to ensure its continued efficiency, longevity, and the delivery of clean, sustainable water to your homestead. This section provides a comprehensive guide to ongoing maintenance and upkeep, covering key components and recommended practices.

1. Scheduled Inspections:

Establish a routine schedule for inspecting all components of your rainwater harvesting system. Regular inspections allow for the early detection of potential issues and ensure timely maintenance.

2. Collection Surfaces:

- **Clean Surfaces:** Regularly clean the collection surfaces, such as rooftops, to prevent debris and contaminants from entering the system.

- Trim Overhanging Branches: Trim branches near collection surfaces to minimize the accumulation of leaves and twigs.

3. Gutters and Downspouts:

- Clear Debris: Regularly clear gutters and downspouts to prevent clogs and ensure unobstructed water flow.
- Inspect and Repair: Check for any damage to gutters and downspouts and repair or replace as needed.

4. First Flush Diverters:

- Regular Checks: Ensure that first flush diverters are functioning correctly by conducting regular checks.
- Clean Mechanisms: Clean the diverters to remove accumulated debris and maintain their effectiveness.

5. Filtration Systems:

- **Inspect Filters:** Regularly inspect and clean filters to prevent clogs and maintain water quality.
- **Replace Filters:** Replace filters according to the manufacturer's recommendations or when they show signs of wear.

6. Pipes and Conduits:

- **Check for Leaks:** Inspect pipes and conduits for leaks and repair any issues promptly.
- **Ensure Proper Slope:** Confirm that pipes have the correct slope to facilitate efficient water flow.

7. Storage Tanks:

- **Inspect Tank Integrity:** Regularly inspect storage tanks for cracks, damage, or signs of deterioration.
- **Check Connections:** Ensure that inlet and outlet connections are secure and watertight.
- **Clean Tanks:** Periodically clean the interior of tanks to prevent sediment buildup.

8. Overflow Systems:

- **Ensure Functionality:** Confirm that overflow systems are functioning correctly to manage excess water during heavy rainfall.
- **Check Pipes:** Inspect overflow pipes for any blockages or damage.

9. Water Level Indicators:

- **Calibrate Indicators:** Calibrate water level indicators regularly to ensure accurate readings.
- **Check for Damage:** Inspect indicators for any damage or malfunctions.

10. Winterization Measures:

- **Drain the System:** In regions with cold temperatures, implement winterization measures by draining the system to prevent freezing.
- **Insulate Pipes:** Insulate pipes and conduits to minimize the risk of freezing.

11. Community Engagement:

- Educational Initiatives: Continue engaging with the community through educational initiatives to promote awareness and responsible water usage.

- Address Concerns: Address any concerns or inquiries from neighbors or local authorities promptly.

12. Record-Keeping:

- Maintain Records: Keep detailed records of maintenance activities, inspections, and any repairs undertaken.

- Document System Changes: Record any modifications or improvements made to the system over time.

By incorporating these maintenance practices into your routine, you can ensure the continuous functionality and sustainability of your rainwater harvesting system. Regular attention to each component, coupled with proactive community engagement, contributes to the long-term success of your water management efforts.

Regular Inspections and Cleaning

Regular inspections and cleaning are fundamental aspects of maintaining a reliable and efficient rainwater harvesting system. By proactively monitoring key components and addressing issues promptly, you can extend the life of your system and ensure the delivery of high-quality water to your homestead. This section details the essential steps for conducting regular inspections and cleaning routines.

1. Scheduled Inspections:

Establish a consistent schedule for thorough inspections to identify potential issues before they escalate. Regular inspections allow you to address maintenance needs promptly, preventing more significant problems and ensuring the optimal functioning of your system.

2. Collection Surfaces:

- Clean Surfaces: Regularly clean the collection surfaces, such as rooftops, to remove debris, leaves, and other contaminants. This prevents the introduction of pollutants into your rainwater harvesting system.

- Check for Damage: Inspect collection surfaces for any signs of damage, such as cracks or loose materials. Address these issues promptly to maintain the integrity of the system.

3. Gutters and Downspouts:

- Clear Debris: Perform regular cleaning of gutters and downspouts to prevent clogs. Remove leaves, twigs, and other debris to ensure unobstructed water flow.

- Inspect for Damage: Check for any damage or corrosion in gutters and downspouts. Repair or replace damaged sections to prevent leaks.

4. First Flush Diverters:

- Regular Checks: Periodically check the first flush diverters to ensure they are functioning correctly. Clean any accumulated debris or pollutants that may affect their performance.

5. Filtration Systems:

- Inspect Filters: Regularly inspect filters for signs of clogging or wear. Clean or replace filters based on the manufacturer's recommendations to maintain water quality.

6. Pipes and Conduits:

- Check for Leaks: Inspect pipes and conduits for leaks. Address any leaks promptly to prevent water loss and potential damage to the system.

- Ensure Proper Slope: Confirm that pipes have the correct slope for efficient water flow. Adjust if necessary to avoid stagnation or reduced flow.

7. Storage Tanks:

- Inspect Tank Integrity: Regularly inspect storage tanks for cracks, damage, or signs of wear. Ensure that tanks are securely anchored and resting on stable foundations.

- Check Connections: Examine inlet and outlet connections to verify their integrity. Repair or replace any compromised components.

8. Overflow Systems:

- Ensure Functionality: Confirm that overflow systems are functioning correctly to manage excess water during heavy rainfall. Check for any blockages in overflow pipes.

9. Water Level Indicators:

- Calibrate Indicators: Calibrate water level indicators regularly to ensure accurate readings. Adjust as needed to reflect the actual water level in the tanks.

- Check for Damage: Inspect indicators for any damage or malfunctions. Replace damaged components for reliable performance.

10. Winterization Measures:

- Drain the System: In areas with cold temperatures, implement winterization measures by draining the system to prevent freezing. Insulate pipes to minimize the risk of damage.

11. Community Engagement:

- Educational Initiatives: Continue engaging with the community through educational initiatives. Promote responsible water usage and address any questions or concerns.

12. Record-Keeping:

- Maintain Records: Keep detailed records of maintenance activities, inspections, and any repairs undertaken. Document any changes or improvements made to the system over time.

By incorporating regular inspections and cleaning into your maintenance routine, you not only ensure the longevity of your rainwater harvesting system but also contribute to sustained water quality and efficiency. These proactive measures play a crucial role in the ongoing success of your sustainable water solution.

Winterization Strategies

Winter presents unique challenges for rainwater harvesting systems, especially in regions with cold temperatures. Proper winterization is crucial to protect your system from potential damage due to freezing conditions. This section outlines effective winterization strategies to safeguard your rainwater harvesting system during the colder months.

1. Drain the System:

Before temperatures drop significantly, drain your entire rainwater harvesting system. This includes emptying storage tanks, conduits, and pipes. This prevents water from freezing inside the system, which can lead to pipe bursts and other damages.

2. Disconnect Hoses and Downspouts:

Disconnect any hoses or downspouts attached to the system. This ensures that there are no residual water pockets that could freeze and cause blockages or damage.

3. Insulate Exposed Pipes:

Insulate above-ground pipes and conduits to protect them from freezing temperatures. Use pipe insulation sleeves or wrapping materials to minimize heat transfer and maintain a stable temperature.

4. Utilize Heat Tape:

Consider using heat tape on vulnerable components, such as pipes and fittings. Heat tape provides a controlled source of warmth to prevent freezing. Ensure it is installed according to the manufacturer's guidelines.

5. Protect Outdoor Components:

Shield outdoor components, such as first flush diverters and filtration systems, from the elements. Use covers or enclosures to prevent snow and ice accumulation, which can affect their functionality.

6. Implement Tank Insulation:

Insulate storage tanks to reduce the risk of freezing. Tank insulation jackets or blankets are effective in maintaining a more stable internal temperature. Ensure that the insulation material is appropriate for your tank type.

7. Install Freeze Protection Devices:

Consider installing freeze protection devices, such as electric heaters or tank warmers, especially in areas where prolonged freezing conditions are expected. These devices help maintain a consistent temperature within the system.

8. Regularly Monitor Weather Conditions:

Stay informed about upcoming weather conditions, particularly freezing temperatures. Plan your winterization efforts based on weather forecasts to ensure timely protection of your system.

9. Perform Visual Inspections:

Regularly perform visual inspections during the winter months. Look for signs of ice formation, especially in pipes and conduits. If ice is detected, take corrective action promptly to prevent further damage.

10. Maintain Records of Winterization Efforts:

Document the winterization measures undertaken each year. This record-keeping helps track the effectiveness of your strategies and informs adjustments for future winters.

11. Educate Community Members:

Engage with your community and share information about winterization practices. Encourage neighbors to take similar precautions to protect their own rainwater harvesting systems.

12. Plan for Spring Reactivation:

As winter ends, plan for the reactivation of your system in the spring. This may involve flushing out any antifreeze used in the system, checking for potential damage, and ensuring all components are ready for the upcoming rainy season.

By implementing these winterization strategies, you not only protect your rainwater harvesting system from potential damage but also ensure its longevity and reliability. Proactive winterization measures contribute to the sustained success of your water management efforts, allowing your system to efficiently resume operations when warmer weather returns.

Repairing and Upgrading Your System

Regular maintenance is crucial for preserving the functionality of your rainwater harvesting system, but occasional repairs and upgrades may become necessary to address wear and tear, changes in water needs, or advancements in technology. This section guides you through the process of identifying issues, conducting repairs, and implementing upgrades to ensure the ongoing efficiency of your system.

1. Identify Issues through Inspections:

Regular inspections play a pivotal role in identifying potential issues within your rainwater harvesting system. Look for signs of leaks, cracks, or component malfunctions during scheduled check-ups.

2. Addressing Common Issues:

- Leaking Pipes or Tanks: If leaks are detected, promptly repair or replace the damaged section. Use appropriate sealants or materials to ensure watertight connections.

- Clogged Filters or Conduits: Clear clogs in filters or conduits to restore optimal water flow. Regularly clean and maintain filtration systems to prevent recurring issues.

3. Upgrading Filtration Systems:

- Assess Water Quality Needs: If your water quality needs have changed or evolved, consider upgrading your filtration systems to meet the new requirements. Advanced filtration technologies may provide improved purification.

4. Expansion for Increased Capacity:

- Evaluate Water Demands: If your water demands have increased, consider expanding your system's capacity. This may involve adding more collection surfaces, larger storage tanks, or additional conduits.

5. Repairing or Replacing Storage Tanks:

- **Cracked or Damaged Tanks:** If storage tanks show signs of cracks or damage, assess whether repairs are feasible. If not, consider replacing the damaged tank with a new one.

6. Implementing Energy-Efficient Components:

- **Upgrade to Energy-Efficient Devices:** If applicable, consider upgrading to energy-efficient components, such as solar-powered pumps or sensors. This not only reduces environmental impact but can also lead to cost savings.

7. Community Collaboration:

- **Engage with Local Experts:** Collaborate with local experts or professionals to assess the overall health of your system. They can provide insights into potential upgrades or improvements based on local conditions.

8. Record-Keeping:

- **Document Repairs and Upgrades:** Maintain detailed records of any repairs or upgrades performed on your rainwater harvesting system. This documentation aids in tracking the system's history and informs future maintenance efforts.

9. Cost-Benefit Analysis:

- **Evaluate Costs and Benefits:** Before initiating upgrades, conduct a cost-benefit analysis. Assess whether the proposed improvements align with the overall goals of your rainwater harvesting system.

10. Stay Informed about Technological Advancements:

- **Keep Up with Innovations:** Stay informed about advancements in rainwater harvesting technology. New innovations may offer more efficient components or systems that could benefit your homestead.

11. Seek Professional Guidance:

- **Consult with Experts:** If in doubt about the best course of action, seek guidance from professionals specializing in rainwater harvesting systems. They can provide valuable insights and recommendations.

12. Community Education on Upgrades:

- **Educate Your Community:** If you implement significant upgrades, share information with your community. Educate neighbors about the benefits and potential changes in water usage practices.

By proactively addressing repairs and upgrades, you contribute to the long-term sustainability of your rainwater harvesting system. A well-maintained and occasionally upgraded system ensures that you continue to enjoy the benefits of sustainable water management on your homestead.

Chapter 6: Maximizing the Benefits

Maximizing the benefits of your rainwater harvesting system involves optimizing its efficiency, adapting to changing needs, and promoting sustainable water practices within your community. This section explores strategies to extract the utmost advantages from your system, ensuring it becomes an integral part of your homestead's water management strategy.

1. Water Conservation Practices:

Encourage water conservation practices within your household and community. Implementing simple measures like fixing leaks, using water-efficient appliances, and practicing mindful water use can significantly enhance the effectiveness of your rainwater harvesting system.

2. Diversify Water Usage:

Explore various applications for harvested rainwater. Beyond traditional uses like irrigation and non-potable water needs, consider integrating rainwater into other processes, such as livestock watering, laundry, or even supplementing your household's potable water needs with proper filtration.

3. Seasonal Planning:

Adapt your water usage and storage strategies based on seasonal variations. During rainy seasons, maximize collection and storage, while in drier periods, implement water-saving measures. This flexible approach ensures optimal utilization throughout the year.

4. Regular System Audits:

Conduct regular audits of your rainwater harvesting system to identify potential areas for improvement. Assess the efficiency of components, the quality of harvested water, and the overall system performance. Address any issues promptly to maintain peak functionality.

5. Community Outreach and Education:

Extend your knowledge and experience by engaging with your community. Host workshops, share success stories, and educate others about the benefits of rainwater harvesting. Fostering a sense of collective responsibility can lead to widespread adoption of sustainable water practices.

6. Integration with Other Sustainable Practices:

Explore synergies with other sustainable practices on your homestead. For instance, consider integrating rainwater harvesting with permaculture principles, greywater systems, or renewable energy sources. This holistic approach enhances overall sustainability.

7. Technology Integration:

Stay abreast of technological advancements in rainwater harvesting. Consider integrating smart technologies, such as sensors for real-time monitoring, automated filtration systems, or apps that provide insights into water usage patterns. Technological enhancements can further optimize your system's efficiency.

8. Harvesting Runoff from Impermeable Surfaces:

Identify impermeable surfaces in your homestead, such as driveways or patios, and explore ways to harvest runoff from these areas. This expands your collection potential and reduces runoff, contributing to both water conservation and improved system performance.

9. Collaborate with Local Authorities:

Engage with local authorities and water management agencies. Understand regional water regulations and explore potential collaborations for community water projects. This collaboration can lead to shared resources, expertise, and collective efforts toward sustainable water management.

10. Continuous Learning and Improvement:

Stay committed to continuous learning and improvement. Attend workshops, read literature on rainwater harvesting, and remain open to innovative ideas. A mindset of ongoing improvement ensures that your system evolves with changing needs and advancements in water management.

11. Evaluate Financial Incentives:

Research and take advantage of any available financial incentives or rebates for rainwater harvesting systems. Some regions offer support or incentives for adopting sustainable water practices, making it financially advantageous to invest in your system.

12. Monitor and Share Success Stories:

Regularly monitor the performance of your rainwater harvesting system and share your success stories within your community. Positive experiences and tangible benefits can inspire others to adopt similar practices, fostering a culture of sustainability.

By implementing these strategies, you not only maximize the benefits of your rainwater harvesting system for your homestead but also contribute to a broader culture of sustainable water management within your community. Your commitment to optimizing water use and sharing knowledge can create a ripple effect, positively impacting the environment and promoting a more resilient and water-conscious community.

Innovative Uses for Harvested Rainwater

Beyond traditional applications like irrigation and non-potable water needs, exploring innovative uses for harvested rainwater can unlock additional benefits and enhance the sustainability of your homestead. Consider integrating rainwater into various aspects of your daily life with these creative applications:

1. Sustainable Landscaping:

Use harvested rainwater to create vibrant and sustainable landscaping features. Implement drip irrigation systems for gardens, flower beds, and ornamental plants. This ensures targeted and efficient water distribution, fostering healthy plant growth while conserving water resources.

2. Greenhouse Watering Systems:

Optimize your greenhouse by implementing a rainwater harvesting system specifically tailored for this environment. Rainwater, with its natural purity, can be ideal for watering delicate plants, fostering optimal growing conditions, and reducing the reliance on conventional water sources.

3. Aquaponics and Hydroponics:

Integrate rainwater into aquaponic or hydroponic systems. The nutrient-rich rainwater can serve as an organic and cost-effective alternative for nourishing plants in soil-less growing environments. This approach combines sustainable water use with innovative agricultural practices.

4. Livestock Watering Stations:

Create dedicated watering stations for livestock using harvested rainwater. This not only ensures a clean and reliable water source for animals but also minimizes the environmental impact by reducing reliance on conventional water supplies.

5. Fire Protection Systems:

Enhance the safety of your homestead by integrating harvested rainwater into fire protection systems. Install strategically placed tanks and conduits to provide a supplementary water source for firefighting, especially in regions prone to wildfires.

6. Water Features and Aesthetics:

Utilize harvested rainwater to create decorative water features, such as ponds, fountains, or cascading water walls. These features not only enhance the aesthetics of your homestead but also contribute to a calming and environmentally conscious atmosphere.

7. Cooling Systems:

Incorporate rainwater into cooling systems for various applications. Use it in cooling towers, air conditioning units, or even for cooling outdoor living spaces. This innovative use not only conserves water but also contributes to energy efficiency.

8. Emergency Water Supply:

Designate a portion of your rainwater storage capacity for emergency water supply. With proper filtration and treatment, this stored rainwater can serve as a backup during water shortages or emergencies, providing a reliable and self-sufficient water source.

9. Sustainable Fish Farming:

Integrate harvested rainwater into sustainable fish farming practices. Create aquaculture systems where rainwater serves as a primary or supplementary water source for fish ponds. This approach aligns with eco-friendly and resource-efficient aquaculture principles.

10. Car and Equipment Washing:

Use harvested rainwater for washing cars, agricultural equipment, and other machinery. Rainwater, being naturally soft, reduces the need for harsh detergents and minimizes the environmental impact associated with runoff from conventional cleaning practices.

11. Educational Demonstrations:

Transform your rainwater harvesting system into an educational tool. Host demonstrations and workshops for your community, showcasing innovative uses and inspiring others to adopt similar practices. Education plays a vital role in creating a more water-conscious community.

12. Artistic Installations:

Explore artistic installations that incorporate harvested rainwater. Rain chains, sculpture features, or interactive installations can serve as both functional components of your rainwater harvesting system and artistic expressions of sustainable living.

By embracing innovative uses for harvested rainwater, you not only diversify the benefits of your system but also contribute to a more sustainable and resource-efficient homestead. These creative applications demonstrate the versatility of rainwater harvesting, turning it into a multifaceted solution that goes beyond mere water collection.

Integrating Rainwater Harvesting with Other Sustainable Practices

Enhance the sustainability of your homestead by integrating rainwater harvesting with other eco-friendly practices. This holistic approach not only maximizes the benefits of your rainwater system but also creates a synergistic and environmentally conscious living environment.

1. Permaculture Design:

Align rainwater harvesting with permaculture principles. Design your landscape to capture, store, and utilize rainwater efficiently. Combine rainwater systems with permaculture zones, creating a harmonious and self-sustaining ecosystem.

2. Greywater Recycling:

Integrate rainwater harvesting with greywater recycling systems. Greywater, generated from household activities like laundry and showering, can complement harvested rainwater for non-potable uses, reducing overall water demand.

3. Renewable Energy Integration:

Explore renewable energy sources to power components of your rainwater harvesting system. Solar panels or small-scale wind turbines can provide energy for pumps, filtration systems, and monitoring devices, making your water management approach more sustainable.

4. Sustainable Agriculture Practices:

Combine rainwater harvesting with sustainable agricultural methods. Implement organic farming, crop rotation, and companion planting alongside efficient water use to create a regenerative and environmentally friendly farming system.

5. Eco-Friendly Building Design:

Incorporate rainwater harvesting into eco-friendly building practices. Design structures with green roofs that not only capture rainwater but also contribute to insulation, biodiversity, and energy efficiency.

6. Community-Supported Agriculture (CSA):

Engage in community-supported agriculture and share the benefits of your rainwater system with the community. Collaborate with local farmers, create CSA programs, and promote sustainable agriculture practices using rainwater.

7. Renewable Energy-Powered Pumping Systems:

Opt for renewable energy sources, such as solar or wind, to power water pumps in your rainwater harvesting system. This reduces reliance on grid electricity and contributes to a more sustainable and off-grid water solution.

8. Agroforestry Practices:

Integrate rainwater harvesting with agroforestry. Planting trees strategically can enhance water retention in the soil, prevent erosion, and contribute to overall ecosystem health, complementing the benefits of your rainwater system.

9. Biodiversity Enhancement:

Design your rainwater harvesting landscape to support local biodiversity. Implement native plantings, create wildlife habitats, and ensure that your water management practices contribute positively to the surrounding ecosystem.

10. Natural Pest Control:

Utilize rainwater to create natural pest control solutions. Implement companion planting and natural predator habitats to reduce the need for chemical pesticides, creating a healthier and more balanced ecosystem.

11. Sustainable Waste Management:

Integrate rainwater practices with sustainable waste management. Implement composting systems, reduce single-use plastics, and adopt practices that minimize environmental impact, contributing to a holistic approach to sustainability.

12. Education and Advocacy:

Combine rainwater harvesting with educational initiatives and advocacy efforts. Share your experiences with the community, promote sustainable living practices, and advocate for policies that support water conservation and environmental stewardship.

By intertwining rainwater harvesting with various sustainable practices, you create a comprehensive and regenerative approach to homesteading. This interconnected system not only enhances the benefits of rainwater harvesting but also contributes to a more resilient, environmentally conscious, and sustainable way of life.

Educating and Engaging Your Community

Empowering your community with knowledge about rainwater harvesting fosters a culture of sustainability, encourages responsible water usage, and creates a network of individuals committed to environmental stewardship. This section outlines strategies for educating and engaging your community in rainwater harvesting practices.

1. Workshops and Training Sessions:

Organize workshops and training sessions to educate community members about the benefits and practices of rainwater harvesting. Provide hands-on experiences, demonstrations, and practical guidance to instill a deeper understanding.

2. Informational Materials:

Develop informational materials, such as pamphlets, brochures, and online resources, that explain the importance of rainwater harvesting.

Distribute these materials in community centers, local businesses, and public spaces to reach a broader audience.

3. Demonstration Sites:

Establish demonstration rainwater harvesting sites within the community. These sites serve as living examples, showcasing the components and functionality of a rainwater harvesting system. Encourage community members to visit and learn from these practical installations.

4. School and Community Gardens:

Collaborate with local schools to integrate rainwater harvesting into educational curricula. Establish rainwater systems in school and community gardens, providing students with hands-on experiences while fostering an early appreciation for sustainable water practices.

5. Community Events and Fairs:

Participate in or host community events and fairs to raise awareness about rainwater harvesting. Set up informational booths, conduct presentations, and engage with attendees to share knowledge and answer questions.

6. Social Media Campaigns:

Utilize social media platforms to launch awareness campaigns. Share success stories, practical tips, and engaging content about rainwater harvesting. Encourage community members to share their experiences and questions, fostering a sense of online community.

7. Neighborhood Collaborations:

Initiate collaborations with neighborhood associations and community groups. Organize neighborhood meetings or events to discuss the benefits of rainwater harvesting, share best practices, and collectively address any concerns.

8. Educational Webinars:

Host webinars featuring experts in rainwater harvesting. Cover topics such as system design, maintenance, and innovative uses. Webinars provide an accessible platform for community members to learn from specialists in the field.

9. Community Contests and Challenges:

Launch friendly contests or challenges within the community to promote rainwater harvesting. Examples include water-saving challenges, garden competitions using rainwater, or the design of creative rain barrel decorations.

10. Collaborate with Local Authorities:

Engage with local authorities and municipal bodies to integrate rainwater harvesting into community planning. Advocate for policies that support and incentivize rainwater harvesting practices, creating a more conducive environment for widespread adoption.

11. Networking Events:

Organize networking events that bring together individuals interested in sustainable practices. Provide a platform for community members to share their experiences, exchange ideas, and form supportive networks focused on water conservation.

12. Educational Partnerships:

Establish partnerships with local educational institutions, environmental organizations, and sustainability-focused groups. Collaborate on initiatives that promote rainwater harvesting education, ensuring a broader reach within the community.

By actively engaging and educating your community, you contribute to the broader adoption of rainwater harvesting practices. The collective knowledge and commitment fostered through these initiatives create a community that is not only environmentally aware but also actively participating in sustainable water management practices.

Conclusion

Embarking on the journey of harvesting rainwater for your homestead signifies a commitment to sustainability, self-sufficiency, and responsible water management. Throughout this guide, we've explored the fundamental aspects of building, maintaining, and maximizing the benefits of a rainwater harvesting system. As you conclude this comprehensive resource, it's essential to reflect on the key takeaways.

Building a Foundation:
The guide began by laying the foundation, emphasizing the importance of rainwater harvesting in providing a clean, sustainable water source for your homestead. We delved into the science of rainwater, understanding its unique properties and distinctions from other water sources. Planning your system thoughtfully, considering legal considerations, and assessing your homestead's water needs were critical steps in establishing a robust foundation.

Constructing Your System:

The construction phase guided you through the practical aspects of building your rainwater harvesting system. From choosing collection surfaces and determining storage capacity to understanding legal and regulatory considerations, each step contributed to the creation of a system tailored to your specific needs.

Maintenance and Upkeep:

Recognizing that the journey doesn't end with construction, we explored the vital role of maintenance and upkeep. Regular inspections, cleaning routines, and winterization strategies ensure the longevity and efficiency of your system. We delved into the intricacies of repairing and upgrading components, emphasizing the importance of adapting to changing needs and technological advancements.

Maximizing the Benefits:

Maximizing the benefits of rainwater harvesting goes beyond the functional aspects of the system. Innovative uses, integration with other sustainable practices, and community education emerged as key strategies. Whether you're exploring artistic installations, combining rainwater harvesting with permaculture, or collaborating with local authorities, these approaches enrich the overall impact of your efforts.

Educating and Engaging the Community:
Finally, recognizing that a collective effort amplifies the impact, we explored ways to educate and engage your community. From workshops and informational materials to social media campaigns and neighborhood collaborations, the guide emphasized the significance of sharing knowledge and fostering a community committed to sustainable water practices.

As you conclude this guide and embark on your rainwater harvesting journey, remember that the success of your system extends beyond the physical infrastructure. It lies in the ongoing commitment to learning, adapting, and actively participating in the broader movement toward sustainable living. By harnessing the power of rainwater, you not only secure a valuable resource for your homestead but also contribute to a more resilient and environmentally conscious community. May your rainwater harvesting endeavor be a source of abundance, sustainability, and inspiration for years to come.